	D/		

GARIBALDI SECONDARY SCHOOL
24789 Dewdney Trunk Road
Maple Ridge, B.C. V4R 1X2

Reading Essentials®
in Science

GLOBAL ISSUES

Consumption and Waste

KAREN E. BLEDSOE

PERFECTION LEARNING®

Editorial Director: Susan C. Thies
Editor: Paula J. Reece
Design Director: Randy Messer
Book Design: Lori Gould, Emily J. Greazel
Cover Design: Michael A. Aspengren

A special thanks to the following for his scientific review of the book:
Wayne B. Merkley, Professor of Biology, Drake University

Image Credits:

© Gary D. Landsman/CORBIS: p. 5; © Bettmann/CORBIS: p. 11; © Sally A. Morgan; Ecoscene/CORBIS: p. 21; © George B. Diebold/CORBIS: p. 27 (middle); © AFP/CORBIS: p. 28; © Galen Rowell/CORBIS: p. 30; © Associated Press: p. 39

© Royalty-Free/CORBIS: p. 14 (top); ClipArt.com: pp. 16 (bottom), 19 (bottom); Image Library: p. 27 (top); MapArt: p. 10; PhotoDisc: front cover, pp. 1, 2–3, 6, 7, 8, 9, 17, 18 (top), 25, 31, 32, 34 (bottom), 36, 37, 38, 41, 44–45, 46–47, 48; Photos.com: back cover, pp. 12, 14, 18 (bottom), 19 (top), 27 (bottom), 33 (lower right), 40; Jill Kimpston: p. 42; Perfection Learning Corporation: pp. 15, 16 (top), 20, 22, 23, 33 (upper left, upper right, lower left), 34 (top), 35, 43; Sassamon Trace Golf Course: p. 24

For information, contact
Perfection Learning® Corporation
1000 North Second Avenue, P.O. Box 500
Logan, Iowa 51546-0500.
Phone: 1-800-831-4190
Fax: 1-800-543-2745
perfectionlearning.com

1 2 3 4 5 6 BA 08 07 06 05 04 03
ISBN 0-7891-6121-4

Table of Contents

1. The Hidden Costs of Consumption 4

2. Garbageology:
   The Study of Waste Throughout History. . . . 9

3. Reducing and Recycling 13

4. Burning and Dumping 21

5. Electronic Waste 26

6. Hazardous Waste 30

7. Working Toward a Solution 36

   Internet Connections and Related Readings . . 44

   Glossary. 46

   Index 48

Chapter One

The Hidden Costs of Consumption

Americans live in a society run by **commercialism**. Commercialism is placing too much emphasis on making money. Advertisements surround us—from billboards to magazine ads to pop-up ads on the Internet. As Americans, we are urged to buy as much as possible to keep the economy strong.

How does commercialism affect pollution and **waste**? Manufacturing the products that we buy causes pollution and waste. Industrial chemicals used in manufacturing and dumped over many decades have made areas of land unfit for use. Products become waste when we dispose of them. **Landfills** for waste disposal are filling up, and communities are concerned about what to do with trash.

To better understand the connection between **consumption** and waste, read what happens when a family buys a new car.

Big spenders

In the year 2000, advertisers spent an average of $2190 per household in the United States to advertise their products.

The Smiths buy a car

Mr. and Mrs. Smith want to buy a new car. They consider such factors as price, gas mileage, safety, and comfort. What the Smiths may not see, however, are the hidden costs involved in producing their new car.

The car that the Smiths are considering began as raw materials in the Earth. Metal ores and other minerals had to be mined to make steel for the frame, wheels, engine, and parts of the body. Glass for the windows and wiring for parts of the electrical system also came from mined minerals.

In the United States, about 98 percent of all metal ores are mined using a technique called *strip mining*. Huge machines strip away soil and overlying rock. Then they scoop out ores and carry them away for processing. Of all the material pulled out of a strip mine, only about 5 or 6 percent can be used. The rest is waste. Sometimes the waste is **toxic**, which means it is poisonous or hazardous to one's health.

Mining hazards

Strip mine

In the late 1990s, the walls around a holding pool for toxic waste from a mine in Spain collapsed. Water from the pool flowed into the Guadiamar River. Volunteers hauled away 10 tons of fish killed by the toxic metals in the water.

Metal isn't all that is needed to make the car. Today's cars contain lots of plastic. The upholstery, carpet, belts, parts of the body, and most of the interior of the car are made from plastic. Crude oil, pumped from deposits deep underground, is used to make the plastics. The processes used to refine crude oil to make plastics produce air pollution and require the use of many toxic chemicals.

Once the materials have been **extracted** and made into metals, glass, and plastics, the car must be assembled. This process begins with preassembly, during which the refined materials are made into car parts. These parts are then shipped to the factories where cars are assembled. Manufacturing these parts can be complicated. The dashboard assembly alone may use the metals iron and magnesium and about 15 different kinds of plastics. Making plastic parts often involves the use of toxic chemicals.

Manufacturers also have to dispose of waste materials. The parts have to be packaged so they are protected during shipment and arrive with identifying labels. The parts may travel hundreds of miles by truck before they arrive at the factory. There they are unpacked, and the packaging is thrown away.

Next, the car is assembled. Once the car is finished and the Smiths drive it away, the car continues to produce waste. Mr. and Mrs. Smith have to fill it with gasoline once a week or so. As they drive, the car burns the gasoline and releases exhaust. The oil and other fluids have to be changed every few months. Tires, belts, and other parts wear out and have to be replaced.

Someday, the Smiths will get tired of this car and want a new one. Their old car may be sold as a used car, but eventually it will be too worn-out to run anymore. It may be sent to a junkyard, where workers can pull out usable parts and sell them. Much of the material in the car, however, will become waste that has to be disposed of. Some of the material, such as oil and transmission fluid, is toxic to the environment. Metal can be recycled, but many of the plastics are not recyclable.

The problem of consumption

Humans, like all living things, have to use, or consume, **natural resources** in order to survive. Plants and some animals are able to store materials for future use. But most living things use resources such as food and water as they need them.

Humans consume resources not only for survival but for their own pleasure. Look up from this book for a moment. How many of the things around you are absolutely necessary for survival?

Most of the products we buy are not necessary for us to stay alive. If we only used the things we needed to survive, life would be pretty dull. Humans enjoy having comfortable places to live and work. We also like to be entertained. The problem, though, is that most of us do not realize how much of our natural resources we consume when we buy things. We contribute to the problems of pollution and waste when we buy things that we don't really need.

Of all the environmental damage done during the lifetime of a manufactured product, most is done before the product ever reaches the consumer. About 94 percent of the materials used during production become waste before the product is even manufactured. This means that only about 6 percent of the raw materials extracted from the Earth ever become products we buy from stores.

In this book, you will become aware of the problems surrounding waste and how you can be a part of the solution.

Each American generates an average of four pounds of trash a day!

Chapter Two

Garbageology:
The Study of Waste Throughout History

Because humans have always used natural resources to survive, humans have always produced waste. How human cultures have dealt with waste has not changed much throughout history. Since the earliest times, people have disposed of waste through dumping, burning, and **recycling**. Only recently has technology made these methods more efficient.

Before we can study the history of waste, we need to define what waste is and what kind of waste we're talking about. What people call "trash" or "garbage" is referred to as *solid waste*. Paper, plastic, cardboard, cans, glass jars, food scraps, old appliances, yard trimmings, and almost everything else that ends up on the curb on garbage day is solid waste.

Early waste disposal

Archaeologists love waste. When digging up sites of ancient villages or cities, archaeologists mainly find things people threw away long ago. Ancient campsites often had heaps of waste from food preparation, such as burned bones, seeds, and other scraps. People also threw away broken tools or other items that were worn-out. By studying discarded items, archaeologists can discover how people used these materials.

Archaeologists also learn how people disposed of trash. For example, ruins of houses in ancient Mayan cities reveal that Mayan housekeepers swept trash into the corners of the house and then covered it over with dirt. The dirt and trash were packed down over many years into solid layers. Mayans also hauled waste to open **dumps**. The rotting trash produced flammable gases, which sometimes exploded.

Cities all over the world had serious waste problems in the distant past. During the Middle Ages, the people of Paris, France, dumped their garbage into piles outside the city walls. The piles became so high that it was difficult for soldiers to defend the city.

UKRAINE
CZECH
SLOVAKIA
MOLDOVA
HUNGARY
ROMANIA
CROATIA
BOSNIA
SERBIA
BULGARIA
MACEDONIA
ITALY
ALBANIA
GREECE
TURKEY
Athens

Rats bred in the garbage and carried fleas. Since fleas can carry the bubonic plague, the garbage piles may have contributed to the Black Death outbreak of the 14th century.

What is Black Death?

Bubonic plague is the medical term for a fatal disease carried by fleas. Black Death was the epidemic of bubonic plague that spread throughout Asia and Europe in the 14th century, killing over 50 million people.

The Industrial Revolution

With the invention of steam-powered machines and factories in the 19th century, people began buying more and more. Manufactured goods became much cheaper. People now working in the cities and factories had to buy the things they needed instead of growing or making things themselves. Increased manufacturing also meant more pollution and waste. People in the 19th century disposed of waste by hauling it to dumps or burning it.

Not all waste was disposed of, however. Many poor people in the 19th century earned a small income through recycling. All products were made of either metal or **organic** materials, so nearly everything produced could be recycled. Poor people searched the streets, alleys, and garbage dumps for usable trash. Rag and bone shops bought cotton and linen rags, which papermakers used as fiber for good paper. The shops also bought bones, which were used in paints and inks. Grease could also be collected and sold to make soap and other products. Employers gave their unwanted clothing to their servants. The servants could either wear the clothing themselves or sell it to used clothing stores. Even dog droppings were collected. Tanners used the droppings to produce fine leather.

The 20th century

The invention of plastics led to new kinds of goods for **consumers** in the 20th century. Modern plastics do not break down. They can be used to make inexpensive toys, furniture, fabric, and other items.

The booming economy in the 1950s and the availability of new, cheaper plastics led to vastly increased production and consumption. This led to more waste.

The situation today

Standards of living have been increasing since the 1950s. This means people are buying more products than ever—and producing more waste than ever. The theme seems to be "bigger is better." The average three-bedroom house of the 1950s was about 900 square feet—the same size as a three-car garage today. Since the 1950s, Americans have used more resources than all the people of the world before then. The average person in the United States uses about 20 tons of raw materials each year. Our nation throws away about 17 million cars each year and about 2 million plastic bottles each hour. The aluminum cans that we throw away each year contain enough aluminum to build 6000 DC-10 airplanes.

With this runaway consumerism comes the problem of what to do with the waste. Most people in Western nations suffer from "NIMBY" syndrome— "Not In My Back Yard." People want better garbage disposal facilities, but only if they're built somewhere else.

Chapter Three

Reducing and Recyling

In 1987, the tugboat *Break of Dawn* had an unusual problem. It could not dock. Not in North Carolina. Not in Louisiana. Not in Texas or Florida. Not in Cuba, Belize, or Mexico. No one wanted to accept the cargo on the barge that the *Break of Dawn* was transporting—186 tons of solid waste from Long Island, New York.

The problem began when Long Island officials passed a law requiring all garbage to be recycled, **incinerated**, or hauled off the island. All of these options were more expensive than using landfills in Long Island. But Long Island's landfills were nearly full.

Alabama businessman Lowell Harrelson came up with an idea. He would load the garbage from Long Island onto barges and ship it to cheap dumps in the South. The barge pushed by *The Break of Dawn* was the first shipment. When the barge pulled into a port in North Carolina, two women took one look at its cargo and called the mayor to complain. The mayor called the governor, and the governor barred the barge from landing in any North Carolina port. This set off a chain reaction. No other port would accept the barge, even those in cities that had signed agreements with Harrelson. The *Break of Dawn* traveled 6000 miles before Long Island officials agreed to let it return home, where the garbage was incinerated.

The question remains—how do we deal with all of America's waste? Experts agree that there are four main ways to tackle the **waste stream** problem. Americans can

- reduce the amount of waste produced,
- recycle,
- burn waste to produce electricity, and
- bury waste in landfills.

Please Recycle

Source reduction

Reducing the amount of waste produced is called *source reduction*. This means consuming less and throwing away less. But it also means manufacturers decreasing the amount of materials or energy used to produce and distribute products. Waste does not only come from items that have been thrown away. It is also generated through the "birth" of a product. It is common sense that the best way to solve the waste problem is to stop waste before it starts.

Reducing packaging

Packaging is what products that we buy are wrapped or boxed in. It serves a purpose. Packaging keeps foods fresh, protects products, and makes shipping and storing of products easier. But sometimes the only purpose packaging serves is to attract consumers to a product so they will buy it.

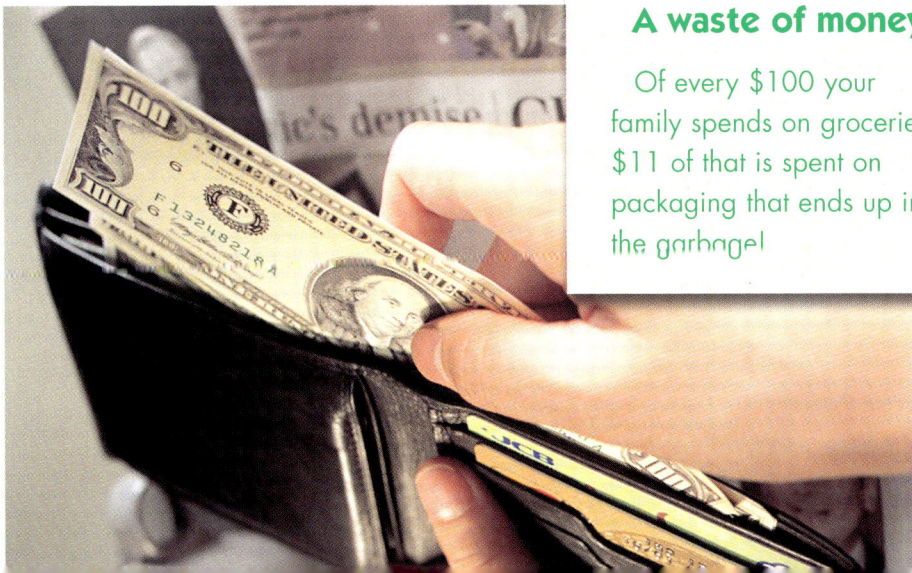

A waste of money

Of every $100 your family spends on groceries, $11 of that is spent on packaging that ends up in the garbage!

Many manufacturers are trying to reduce the amount of waste they produce. Packaging makes up one-third of all garbage in the waste stream in this country. One way to reduce packaging and waste is by offering consumers refills. Concentrated products, also known as "ultras" in the cleaning industry, require less packaging—up to 75 percent less! Consumers can purchase containers made of paperboard so they don't have to buy plastic jugs every time. Also, not as much energy is needed to ship smaller containers either.

Paperboard refill containers require less packaging than plastic jugs.

Companies are also looking at the way packages are designed to find ways to reduce waste. Nearly all bottles, cans, and jars are produced with fewer materials today than they were 30 years ago. The weight of one plastic 2-liter soft drink bottle has been reduced from 68 grams in 1977 to 51 grams today. What difference does a few grams make? Would you believe this change has kept 250 million pounds of plastic out of the waste stream every year?

McSource reduction

McDonald's has been a leader in waste reduction in the fast-food industry, winning the WasteWise Partner of the Year award in 2000 given by the U.S. Environmental Protection Agency. McDonald's redesigned its straws, napkins, sandwich packaging, cups, French fry containers, and other items to reduce the packaging. This fast-food giant has saved 150,000 tons of paper and cardboard from the waste stream over the past ten years.

Reducing number of purchases

Source reduction not only refers to the manufacturing of products, but also the consumption of products. Americans today are busier than ever. They want convenience, and many times products that are convenient require more packaging and are not necessary. These products are purchased to make life easier. Purchasing for convenience only produces more waste.

Recycling

Studies have shown that 2 to 5 percent of the waste stream can be used again. Using something again is known as *recycling*. Americans are making recycling a part of their routine. In 1960, only 6 percent of trash was recycled. Today, Americans recycle 28 percent. In 1999, 64 million tons of trash that would have gone to landfills was recycled.

Why recycle?

Using something again eliminates that item from becoming waste. Recycling can be selling used clothing at a garage sale or giving a CD you don't want anymore to a friend who would like it. Recycling is also sending products to recycling centers, where they will get converted back into raw materials. New products can then be made using these materials. This most often requires less energy than manufacturing products from new materials. For example, making new aluminum cans from recycled aluminum uses 96 percent less energy than making cans from all-new materials.

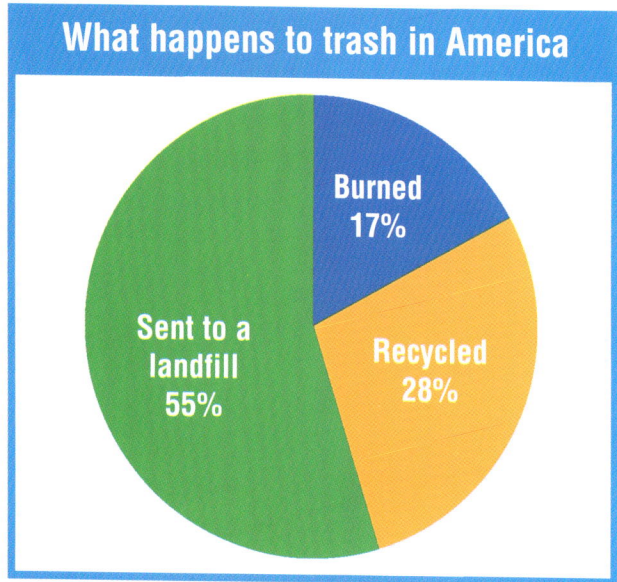

What happens to trash in America

Burned 17%

Recycled 28%

Sent to a landfill 55%

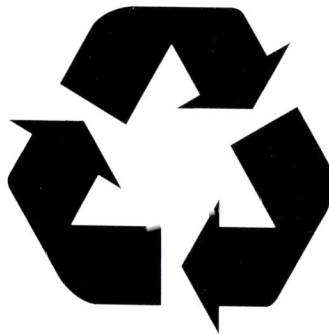

Recyclable
This symbol means the materials in the product can be recycled into a new product after it is used.

Recycled
This symbol means that at least part of the material used in the product was used before.

Recycling paper

What do you think makes up the biggest portion of the waste stream? Styrofoam? Diapers? Actually, the correct answer is paper, which makes up 39 percent of the waste stream. With all of that paper being wasted every year, recycling becomes very important.

Recycled paper can become many products. Old newspapers can be made into new newsprint, egg cartons, and paperboard, such as what cereal boxes are made of. Discarded corrugated boxes can become new boxes or paperboard. And white paper, the kind used in printers or copy machines, can be made into almost any paper product.

Purchasing recycled paper and paper products also supports recycling. Recycled paper is made from mixing waste paper and fresh wood pulp.

Recycling glass

Paper can be either recycled or burned to produce energy. Since glass doesn't produce energy when burned, the best disposal method for it is recycling. Today, 40 percent of the glass that is disposed of in the United States is recycled.

Glass cannot produce energy, but recycling glass saves energy. It takes 30 percent less energy to make glass from recycled glass than to make it from new materials.

Glass
- Separate by color / Remove Lids.
- Keep Breakage to minimum.
- Container Glass only.
  (No Window Glass, Please.)

Brown Glass

Green Glass

Clear Glass

Recycling metals

Every day, Americans use and dispose of 100 million steel cans and 200 million aluminum cans, which contain beverages. Think about how many cans of food are in your pantry.

Like glass, metals don't produce energy when burned. But they do save energy when recycled. It takes only one-fourth of the energy to make steel from recycled steel than from raw materials.

Aluminum cans are easy to produce from recycled aluminum—an example of "closed loop recycling."

Why are steel cans called "tin cans"?

Steel cans usually have a thin coating of tin that protects the cooked food in the can.

Recycling plastics

Plastics are everywhere. Look around the room. How many things do you see that contain plastic? There are more than 10,000 different types of plastic. But only 5 percent of the plastics in America are recycled!

Plastics sometimes get a bad rap. But the truth is that they are really energy-efficient. Plastics can be burned to produce energy or be recycled.

Recycling some types of plastic is difficult. It actually takes more energy to make products from some recycled plastic than from all-new materials. However, some plastics are very recyclable. One example is soft-drink bottles. Once melted down, this plastic can be used to make carpet, T-shirts, stuffing for jackets, or even new soft-drink bottles.

Plastics in packaging

Over 40 percent of all plastics are used for packaging!

Recyclable Plastics

1 PETE	Polyethylene Terephthalate	Two-liter beverage bottles, mouthwash bottles
2 HDPE	High Density Polyethylene	Milk jugs, trash bags, detergent bottles
3 V	Vinyl	Cooking oil bottles, packaging around meat
4 LDPE	Low Density Polyethylene	Grocery bags, produce bags, food wrap, bread bags
5 PP	Polypropylene	Margarine tubs, diapers
6 PS	Polystyrene	Hot beverage cups, take-home boxes, egg cartons, meat trays
7 Other	Other	All types of plastics or packaging made from more than one type of plastic

Activity

Fast trash

How much trash do you create when you eat at fast-food restaurants? Get several friends together for this activity. Have each person in your group choose a different restaurant and buy a complete meal. Agree to all meet somewhere to eat. When you're done, pile up the trash. Which restaurant gave you the most trash? How much of it can be recycled? Did any of the restaurants provide bins for recycling glass or metal?

Chapter Four

Burning and Dumping

Turning waste into energy

Before the 1970s, many people in the United States burned their own household trash in backyard burn piles, burning barrels, or household incinerators. However, open burning was a source of pollution. Most communities in this country have banned open backyard burning within city limits. Burning solid waste is now done in **municipal** garbage incinerators.

Burning solid waste has one major advantage—it reduces the amount of waste by 70 to 90 percent. This greatly reduces the amount of waste that goes to a landfill. Today's municipal garbage incinerators are designed to release far fewer pollutants than backyard burning. Close to 20 percent of all waste in the United States is incinerated.

This incinerator in Long Island, New York, burns all types of waste to generate electricity.

How incinerators work

A typical garbage incinerator collects solid waste in a concrete-lined pit. A crane operator uses a large claw to mix the garbage and remove large, nonburnable items such as appliances. The operator then scoops up the garbage and dumps it into a chute, where it is carried to a chamber and incinerated at a temperature equal to that of molten lava from a volcano. Heavy ash, called *bottom ash*, settles to the bottom of the chamber where it can be removed. Magnets are used to remove scrap iron and steel, which is then recycled. Lighter ash, called *fly ash*, often rises into the smokestacks, where it is collected in filters.

The heat produced by burning trash can be used to create electricity. Paper, plastics, wood, and packaging all make excellent fuels when burned. **Waste-to-energy plants** produce enough electricity to supply nearly 2.5 million households today.

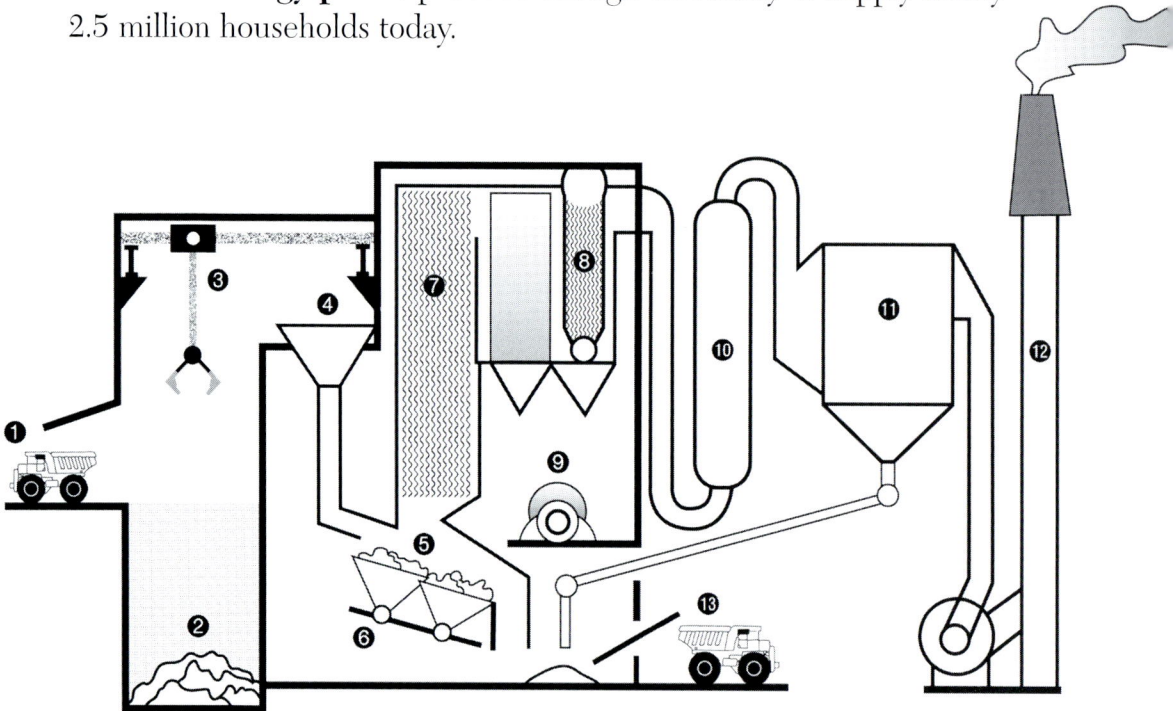

① Tipping area for trucks ② Refuse pit ③ Refuse crane ④ Chute ⑤ Primary combustion zone
⑥ Underfire air ⑦ Furnace ⑧ Heat exchanger ⑨ Turbine ⑩ Scrubber, to remove acid gases
⑪ Fly ash and dust collector ⑫ Stack ⑬ Bottom ash and fly ash collection and transport

Landfills

Not all waste can be recycled or burned. So what's the alternative? Today, about 55 percent of the solid waste in the United States is disposed of in landfills, which are quite different from the old-fashioned open dump.

What a dump!

A dump is simply a pit in the ground where people throw their garbage. Some of the trash decomposes, or breaks down, over time. Garbage near the surface rots, creating foul odors. Open dumps attract rodents and other pests. Rotting garbage produces **methane**, a gas that contributes to the warming of the atmosphere. Methane is also flammable. It's not uncommon for open dumps to burn or even explode.

Chemicals leaking from open dumps are also a problem. As rain falls on the dump, chemicals may leak out of the trash, forming a solution called *leachate*. The leachate may find its way into groundwater or streams, causing serious water-pollution problems.

How a landfill is made

The landfill is designed to reduce the problems of open dumps. The modern landfill is lined with a layer of clay and may have a plastic liner as well. This isolates the trash from soil and groundwater. The lined unit is divided into cells, which are filled one day at a time.

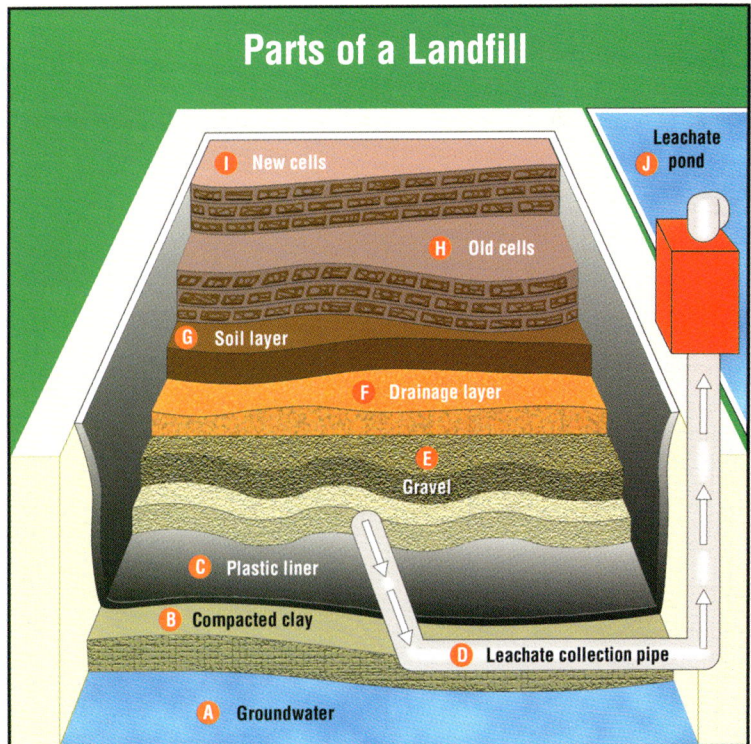

Parts of a Landfill

I New cells
H Old cells
G Soil layer
F Drainage layer
E Gravel
C Plastic liner
B Compacted clay
D Leachate collection pipe
A Groundwater
J Leachate pond

Compacting garbage limits the oxygen that reaches it. The lack of oxygen actually preserves garbage rather than allowing it to decompose, like in a dump. Scientists studying old landfills have found readable newspapers that are 30 years old!

At the end of the day, each cell is covered with 6 inches of soil and compacted more. The layer of earth put on a cell is called "the daily cover." This reduces odors and pests. As trash is added, heavy machines compact it.

A drainage system at the bottom of the cell carries the leachate away to be processed. Methane gas is given off by organic materials as they rot. This is often collected and used as fuel, known as *natural gas*.

Once the landfill is full, it is capped with a thick layer of soil. Vegetation is planted to keep the soil from wearing away. Old landfills may become new parks, ballfields, ski slopes, golf courses, or wildlife habitats.

The major problem with landfills, however, is that while most people understand that they are necessary, no one wants to live near one. This makes finding a site for a new landfill a major challenge.

Sassamon Trace Golf Course in Natick, Massachusetts, is a nine-hole course built in 1999. Four of the golf holes rest on top of an old landfill. The cap of the landfill was shaped with the intention of having golf holes on it.

Activity

Dump vs. landfill

You can model the difference between an open dump and a sanitary landfill with some simple materials. Start with two plastic colanders or old sieves. Line one with a thin layer of modeling clay to represent the liner of a landfill. Leave the other unlined. Tear up some red tissue paper or crepe paper, and mix it with household garbage, such as vegetable scraps, torn paper, cardboard, and scraps of plastic containers. Put a layer of garbage about one inch thick in both colanders, and cover with a half-inch of potting soil. Continue layering until both colanders are full. Set the colanders on shallow pans to catch the leachate.

Pour one cup of water into each colander twice a week for four weeks. Observe any leachate that seeps through. The red tissue paper will color the leachate, representing toxic chemicals that could seep out of a dump or landfill. Which would be better for the environment?

Chapter Five

Electronic Waste

Solid waste isn't the only type of waste that we create. Every year, we see advances in technology and industry. Think about the technology you have at home or at school. How many electronic devices can you think of? Moving ahead, however, has its price. The modern world has created new waste problems that need to be solved.

Precious waste

In a village in southeast China, a strange scene is playing out. Villagers who have scavenged old computer parts from a nearby recycling center are busy burning some parts and pouring acid over others. Clouds of toxic smoke and fumes rise into the air. What are these villagers doing? They are attempting to extract the small amounts of gold and silver that were used to build the computer parts. They hope to sell these precious metals. Their extraction techniques, however, endanger their health as they use harmful acids and release toxic gases.

This village isn't an unusual case. About 80 percent of all discarded electronics collected for recycling in Western nations are exported before they are recycled. The majority end up in China.

Outdated already?

You've probably heard people say, "I just bought this computer, and it's already out of date!" You or one of your friends may be wishing for the latest, greatest video game system. Or maybe you want to replace an old VCR with a brand-new DVD player.

The problem with most electronics is that they become outdated so quickly. People want the latest computers, the smallest cell phones, or the video game system that plays all the newest games. By producing newer and better products every few months, electronics manufacturers can sell more products and make more money. Software writers create bigger programs and games that require more powerful systems. This almost forces people to replace their electronics frequently just to stay current, which results in an incredible amount of waste.

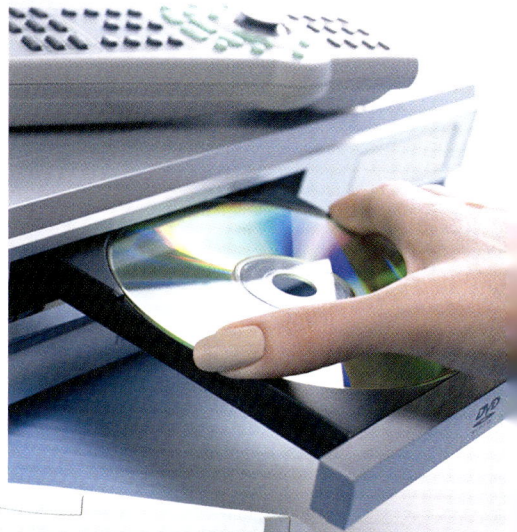

Where have all the old computers gone?

In 1998, more than 20 million personal computers became outdated. However, only 14 percent were reused or recycled, which means that the rest either ended up in landfills or are still lurking in storage.

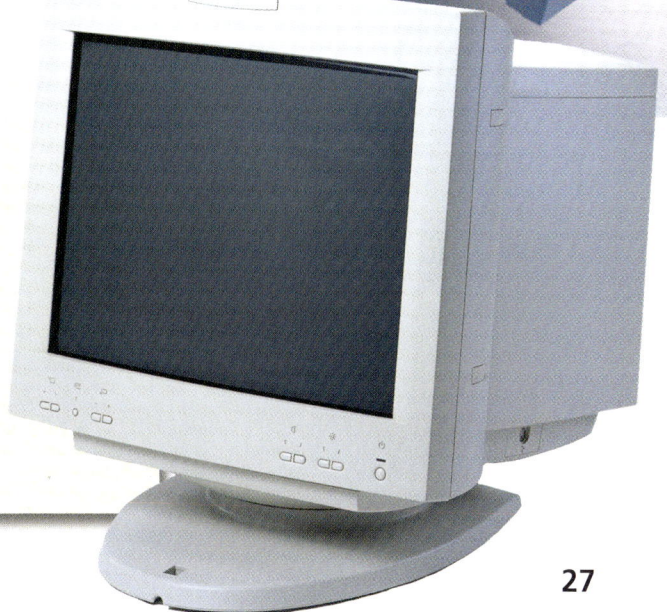

The electronics waste stream

Many different materials are used to produce electronic devices. Any one device may have metal parts made from aluminum, titanium, copper, and steel. A dozen different plastics may be needed. Manufacturing the various plastic parts uses toxic chemicals and large amounts of petroleum, which could hurt the environment.

Most companies that make computer chips use chemicals called *chlorofluorocarbons* to clean them. These chemicals can damage the atmosphere. Computer-chip manufacturers must be extremely careful to keep their chlorofluorocarbon solutions contained so they do not leak. But it is impossible to prevent all leaks.

Old electronics are extremely difficult to dispose of because they are made of so many different materials. About 70 percent of electronic devices end up in landfills, where their valuable

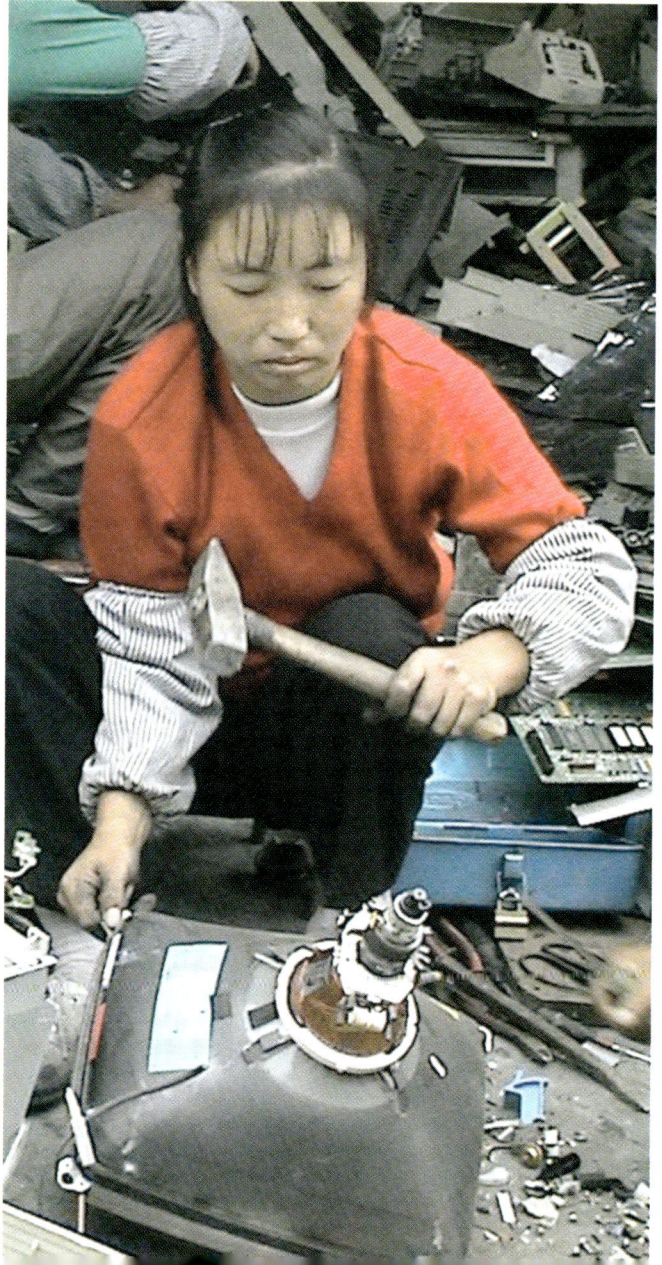

A woman in rural China uses a hammer to retrieve copper from the cathode ray of a computer monitor.

components go to waste. Most electronics produced today are made of recyclable materials, including metals, glass, and recyclable plastics. But separating the materials takes a lot of time and labor. Old electronics may be shipped to other nations where labor is cheaper. However, in places like the village in China, poor people may scavenge the recycling collection centers and try to extract materials themselves using dangerous methods.

Problems in recycling electronics

The first step in recycling electronics is collecting old devices. Most consumers, however, don't know where to take their old electronics. Some communities or industries have special one-day events where people can bring in old computers and electrical appliances. A few communities have curbside pickup service. Because it's hard to find a place to get rid of old devices, many people either let their old electronics sit unused in storage or just dump them into the garbage can.

Once electronics are taken away for recycling, they may be evaluated. Usable electronics may be repaired and sold. The rest are sent to a place where they are taken apart and the materials are separated. This is called *demanufacturing*. However, different plastics require different kinds of recycling processes and must be separated from one another. Very few plastic parts carry labels that identify what they are made out of, so it's difficult for the demanufacturer to separate the plastics.

A third problem is finding a use for recycled plastics. Though many modern plastics can be recycled, there are few manufacturers that need recycled plastic. This makes it difficult for electronics recyclers to resell sorted plastics.

With technology changing every day, **electronic waste** will continue to be a growing problem.

What is recycled the most?

A study of residential electronics collection programs showed that 36 percent of the electronics recycled were TVs, 16 percent were stereo equipment, and 11 percent were computer monitors.

Chapter Six

Hazardous Waste

Children growing up in the 1970s in the Love Canal housing project near Niagara, New York, played some exciting games. They played with bluish rocks that sparked and exploded when they were hurled at a cement wall. They used these in games of pretend war, then led their captives off to cellars where blue-black goo oozed up from cracks.

Unfortunately for these children, their playthings turned out to be strange combinations of toxic industrial waste. From the 1930s until 1952, the Hooker Chemical and Plastics Company had used the old Love Canal bed as an open dump for chemical waste. In 1953, the canal was filled in and capped with clay. A school and a housing development were built on top. Despite the clay cap, chemicals oozed up from the old canal. Children in the

Love Canal is abandoned today.

Over 6.5 billion pounds of toxic chemicals from industries are released into the environment every year.

area had suffered high rates of serious illnesses for several years, including cancer and birth defects, but no one knew why.

Government officials were reluctant to believe that the exploding rocks and blue goo could be the cause. Manufacturers were also nervous about the idea, since their companies produced many toxic wastes and disposing of them safely was expensive. Parents, teachers, and health workers fought for official recognition of the problem and demanded immediate action. Finally investigators tested the ooze and found at least 82 different toxic chemicals. In 1978, the Love Canal housing project was condemned.

What is hazardous waste?

Hazardous waste is any waste substance that poses a risk to human or environmental health. Deliberate dumping of chemical hazardous waste isn't the only way people can be harmed. Chemical accidents can also occur.

Early in the morning on December 4, 1984, many residents of Bhopal, India, woke up to a chemical cloud that caused burning eyes, coughing, nausea, dizziness, and vomiting. The cloud was methyl isocyanate, a poisonous gas that had leaked from a chemical plant. Everyone within a five-mile radius was affected. Most of those caught outdoors within a few miles of the plant died immediately. Over 3500 people were killed, and 200,000 more were injured.

Hazardous waste can hurt you if you eat or drink it, breathe it, or if your skin comes in contact with it. Reactions range from minor irritation with some short-term exposure to death with some long-term exposure.

Over 300 million tons of hazardous chemical waste is produced every year in the United States. Some of these chemicals are produced from the manufacture of plastics. Because people want toys, CDs, radios, computers, clothing, and other manufactured products that use plastics, manufacturers have to make the plastics. In doing so, they produce toxic waste. Other hazardous materials are used in assembling products, including paints and chlorine gas.

POISON

DANGER-CAUSES SEVERE BURNS

READ INSTRUCTIONS ON CARTON AND BOTTLE BEFORE USING THIS

Household hazardous waste

Hazardous waste is not only produced with the manufacture of products. Some hazardous products can be found right in your home. If not used or disposed of properly, these become hazardous waste. They may not only hurt you but also the environment if they seep into the land or water.

Common Household Hazardous Products

In the Kitchen

Scouring powder
Oven cleaner
Glass cleaner
Furniture polish
Drain cleaner

In the Bathroom

Toilet bowl cleaner
Hair spray
Nail polish
Nail polish
    remover

In the Laundry Room

Bleach
Laundry detergent
Spot remover

In the Garage

Antifreeze
Motor oil
Car wax
Paint
Insect spray
Weed killer

How do you know if the products in your house are considered hazardous? It should be pretty easy to tell. The federal government passed a law requiring labels on hazardous substances. A product marked "poison" is of the highest hazard level. This indicates that it is poisonous and can cause injury or death if breathed, eaten, drank, or absorbed through the skin. The next level is "danger," which means it can be poisonous, can catch on fire, or can be corrosive. "Caution" is the least serious level.

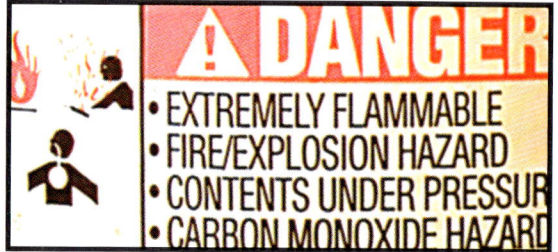

Disposing of hazardous waste

If you use hazardous substances, how do you get rid of them if you haven't used them up? Should you toss them in the trash? Dump them on the ground or down a storm drain? Absolutely not! They should be taken to a hazardous waste collection site, which most communities have. From there, they will be taken to proper hazardous waste landfills.

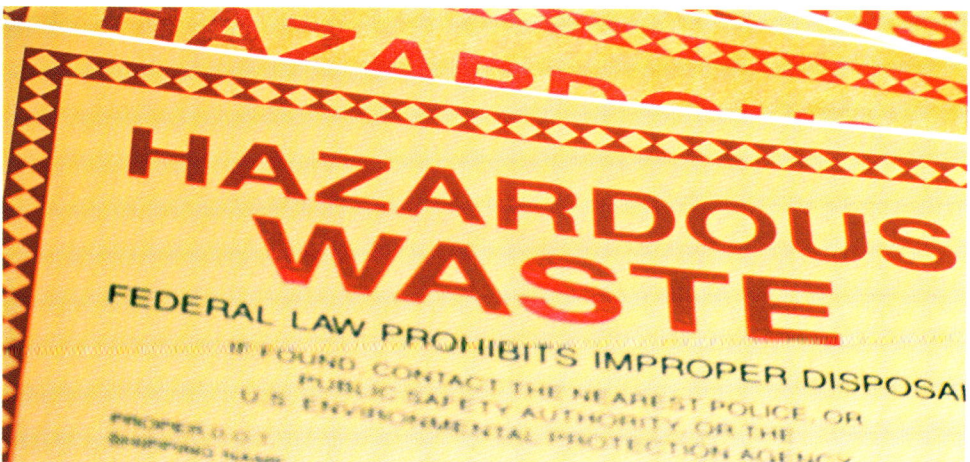

The best way to deal with hazardous waste, however, is to not create it in the first place. When possible, it is beneficial to use alternatives to hazardous chemicals.

Polluting our water

One of the largest sources of water pollution is used motor oil. This pollution comes from people who change the oil in their cars themselves and don't properly dispose of it. Most automotive stores or shops will take used motor oil for recycling or proper disposal. The taste of up to one million gallons of water can be ruined by just one gallon of used oil that has been dumped into it!

Activity

Hazards around the house

You may have hazardous chemicals in your house and not even know it. Do an inventory of the chemicals in your house. Look for cleaning products, air fresheners, paints, garden chemicals, insect sprays, laundry products, and automotive products. Write down what you find. Check the labels for warnings, and find out what kind of hazards are associated with these chemicals. Then check the labels and see if they tell you how to safely dispose of the chemicals and the empty containers. Some will, but many will not. You can call your city or county waste disposal department and find out how to safely dispose of household toxic chemicals.

35

Chapter Seven

Working Toward a Solution

What governments and industries are doing

The federal government and state and local governments are making efforts to address the waste stream problem and work toward a solution. Industries are also doing their part to make changes and encourage waste reduction and proper waste management.

The Superfund Act

In 1980, the federal government passed The Superfund Act to supply money for toxic waste cleanup. The act allows the government to seek payment from manufacturers who dumped the waste in the first place, regardless of how long ago the dumping occurred. Individuals and businesses that generate hazardous waste are required by law to dispose of it safely and are legally responsible for any environmental contamination the waste causes.

The Superfund Act has been responsible for cleaning up over 1300 sites in the United States. These include old factories, landfills with poisonous waste, and old mines with toxic metals.

Local government programs

States, cities, and communities have also come up with programs and laws to try to help the waste problem. Look at the top of one of your soda cans. Does it say "5¢" or "10¢"? If so, your state has a program that pays people for turning in beverage cans or bottles for recycling. This is sometimes called "the bottle bill." State leaders hope that this encourages people to recycle precious aluminum and glass.

Many cities and communities now have voluntary recycling programs. In 1999, around 50 percent of Americans had access to curbside collection programs. This means that residents can put a recycling bin on the curb and have recycled items picked up by a collection service. The convenience of curbside programs makes it more likely that people will recycle.

Some communities have now adopted "pay-as-you-throw" programs. This means that residents are charged for each bag of trash the garbage collectors pick up. This encourages people to use and throw away less. It also helps make them want to recycle what they can.

Along with "pay-as-you-throw" programs, mandatory recycling has been instituted in some communities. If residents throw away

More than 4000 communities have adopted "pay-as-you-throw" programs.

items that are supposed to be recycled, they are given a fine or they will not get their trash picked up. New York City is the only big city to make recycling mandatory. The Sanitation Department actually has officers who go through people's trash to make sure they are recycling! People who break the law are fined $25 each time.

Industry pre-cycling

In addition to recycling, "pre-cycling" helps reduce waste. *Pre-cycling* means reducing the creation of waste during the manufacturing process.

For many years, the cheapest materials for manufacturing were raw materials from the Earth. Because these materials were cheap and abundant, there was no economic reason to reduce waste.

Today, greater consumer demands mean natural resources are being used up rapidly. So, environmental laws have been enacted to control how natural resources are used. In an effort to reduce industrial waste, it's becoming more and more cost-effective to operate "green factories."

Definition of "green"

Everyone knows green is a color. But it is also a term used to mean "environmentally friendly." A "green factory" respects the environment by practicing pollution control and waste reduction.

Pre-cycling involves changing ways in which goods are made to reduce waste during the process. It makes good sense to reduce the amount of materials that end up in a factory's trash bin. All that trash was originally raw material the manufacturer paid for. By reducing the amount of waste, the manufacturer reduces the amount of raw materials needed.

Electronics industry

Many computer manufacturers are making it easier for customers to recycle old computers. Apple, Sony, and Gateway all have programs that allow computer owners to return their old machines to the company for recycling. Most companies now sell computers in parts for easier upgrades, so the entire machine may not have to be replaced. Apple is also actively researching ways to redesign computers to reduce the amount of material used in the first place.

The American Plastics Council is also trying to make electronic recycling easier. The Council is promoting ways to mark plastic parts of electronic equipment to make it easier to identify the type of plastics the parts are made from. The Council is also researching easier ways to identify parts in old equipment.

What manufacturers haven't solved is the problem of electronics becoming outdated so quickly. As long as newer and better products keep coming out every few months, people will continue buying new devices.

Discarded computers are piled up in bins at CP Recovery, an Omaha, Nebraska, company specializing in the recycling of discarded electronics equipment.

What you can do

Reduce demand

When people concerned about the environment point a finger at manufacturers and accuse them of creating a waste problem, manufacturers often reply that they are only meeting consumer demand. They claim they make products only because people want them. Yet industry often creates demand in the first place through advertising. Think of the ads you see for your favorite candy, soft drink, toys, or electronics. Notice how the ads make the products seem fun, exciting, and fashionable. Ads are made to convince people they need these products.

As consumers, we are in charge of what we buy. When we make decisions to reduce our consumption, we also reduce the demands we make on the natural resources of the world. We reduce the amount of waste and pollution involved in manufacturing products.

Of course we have to buy some things. We need food and clothing to survive. We all want nice houses and furniture. Our economy is founded on consumption. Reduced consumption can hurt the economy. We need to be aware, though, of the environmental cost of the things we want.

One easy way to help reduce your consumption is to find noncommercial things to do. Instead of watching TV or playing the latest video game, play a sport or game outside. Instead of collecting the latest trading cards, start another hobby, such as planting a small garden, learning about the stars and planets, or drawing.

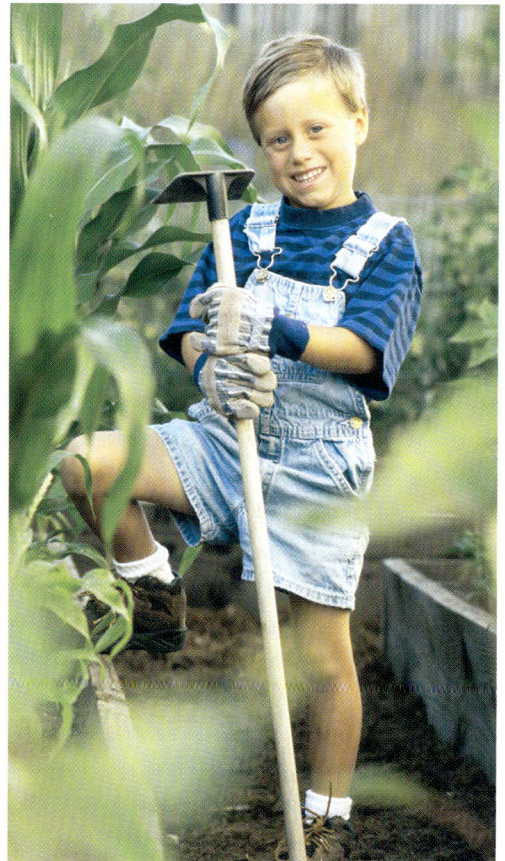

Reduce waste

There are other ways you can contribute to source reduction. Do you use your local or school library for books and information? You should! It's free, and it is based on recycling. Talk to your family about using the library or the Internet for news instead of subscribing to a newspaper and shopping online instead of from catalogs that are sent to your house. Using computers in a library helps cut down on electronic waste also.

Try to reduce what you use. When you buy only one or two things from a store, tell the clerk to keep the bag. Carry the items in your hands or in your own bag. When you can, buy products with less packaging. This includes concentrated products or refills. And buy items that are durable. They may cost a bit more than their competitors, but they will last longer, producing less waste in the long run.

Some stores create their own incentives for reducing waste and recycling.

What to do about unwanted catalogs

Does your family receive catalogs in the mail that you don't want? Sometimes it's hard to imagine why your name is even on certain mailing lists. Just why do you receive a catalog full of nothing but Scottish kilts?! Here's what you can do. Send a letter to the following address, and tell them to save money and waste by taking your name off their mailing list.

Mail Preference Service
Direct Marketing Association
11 W. 42nd St. P.O. Box 3861
New York, NY 10163-3861

Recycle at home

Recycling is an easy way to do your part in reducing waste and saving valuable resources. There are many different ways you can recycle—some may not even seem like recycling!

Garage sales are a good way to recycle clothes and household items.

Do you shop at garage sales? Did you know this is a form of recycling? Just think—when you buy something that someone else doesn't want and you will use, you are eliminating that product from entering the waste stream. You can also shop at Goodwill or other charity stores before buying something new. Of course, holding your own garage sale or donating unwanted items to charity also helps reduce waste.

Does your family use its leftovers, or does that leftover food get tossed into the garbage? Over two-thirds of the food we throw away is edible! Find creative ways to use leftover food to get the most use out of it. This will not only reduce waste but your family's grocery bill too!

The way you can really make a difference is by finding out if your community has a recycling program. If so, recycle! Follow the instructions given by your collection service, and recycle everything you can. If your community doesn't have a curbside service, find out the closest place where you can drop off recyclable items.

Here are some other easy ways to recycle:

- Use your own canvas or cloth bags when you go shopping instead of taking home more plastic or paper bags.
- Use cloth napkins that are washable instead of buying disposable paper napkins.
- Use recycled or recyclable gift wrap. Try cloth bags or the comics section from the newspaper!
- When you go to a fast-food restaurant, take a travel mug for your drink instead of using one of their disposable cups.
- Wash plastic storage bags and utensils and reuse them.

- Use sponges and rags for cleaning instead of paper towels. You can turn old towels and T-shirts into rags instead of throwing them away.
- Reuse margarine tubs and cottage cheese containers to store or freeze food.

Recycle at school

Does your school or classroom participate in recycling? There are some easy ways to start recycling. Think about how much paper you use at school. If your school doesn't recycle paper, ask your teacher or principal if you can start. Remember that paper makes up the biggest percentage of the waste stream! Do you use just one side of a sheet of paper before throwing it away? Ask your teacher if you can use both sides before recycling it. Create a box in your classroom where you can donate and reuse pens, pencils, folders, and anything else that you could rescue from the waste stream.

Think about recycling at lunch too. Take a look at your school lunch program. Do you use disposable eating utensils? Disposable trays? If so, can you recycle them? Ask your teacher or principal if recycling is being considered in the lunchroom. If you take your own lunch, do you have a reusable lunchbox or sack? Look at how much of your lunch is disposable. Can you bring food in recyclable containers instead of disposable bags?

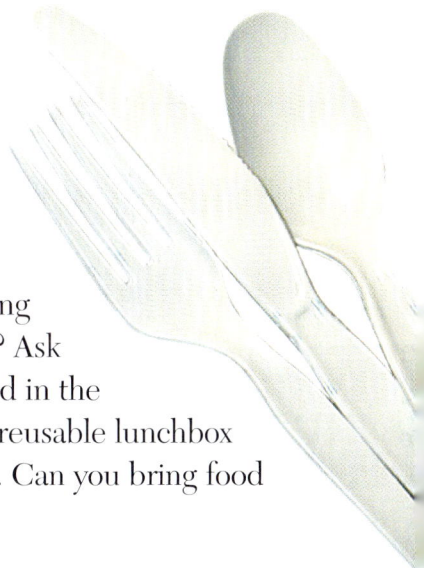

Conclusion

In the 40 years between 1960 and 2000, the amount of waste that Americans produced more than doubled—from 88 million tons to 225 million tons. The more we buy, the more waste we create and the more natural resources we use up to create new products. The waste stream can be managed by using proven methods of disposal such as landfills or incinerators. Recycling eliminates more waste from being produced, and of course, reducing waste at its source by being a smart consumer is the best solution of all!

Internet Connections and Related Readings

Ecological Footprint Quiz (http://www.MyFootprint.org/)
Take this easy quiz to learn what your ecological footprint is in the areas of food, mobility, shelter, and goods and services. How much are your actions affecting the environment? Learn how many "acres" you take up. Find out how many planets we would need if everyone lived like you do.

Learn About Chemicals Around Your House
(http://www.epa.gov/kidshometour/)
In this site, sponsored by the U.S. Environmental Protection Agency, you can learn about pesticides and other toxic substances that are lurking around your home. Click on different parts of a house, and find the household products that contain pesticides or other toxic substances. Click on each to find out what it is and what health and safety information you need to know. Explore a generic label to find out about parts of hazardous labels and the different signal words. Learn what to do if you have an accident with a toxic substance. After you have become a well-informed student of toxic substances, complete a word search, word scramble, or crossword puzzle, or test your knowledge with a quiz. You can even submit your own question!

Planet Protectors Club

(http://www.epa.gov/epaoswer/osw/kids/index.htm)

This site, also sponsored by the U.S. Environmental Protection Agency, explores the problems and solutions of solid waste. Print out activities to complete, such as "Trash and Climate Change." In the activity "The Case of the Broken Loop," follow a detective through activities to learn more about reducing waste and conserving resources. Play games such as "N. Trubble and the Environauts," where you learn about an urgent need to prevent trash from taking over the planet. You must find ways earthlings can reduce the amount of waste they create. In "Which Bin Does It Go In?," test your reflexes and recycling savvy by dropping trash from a conveyer belt to the correct recycling bin before any items fall into the trash can. You can also complete the "Waste No Words Crossword Puzzle." Finally, read the comic book "Adventures of the Garbage Gremlin," and find out what different things you can do with a used jelly jar.

Recycle! by Gail Gibbons. A step-by-step tour through the recycling process. Includes suggestions for starting your own recycling program. Little Brown, 1996. [RL 2 IL K–3] (4965202 CC 4965201 PB)

Where Does the Garbage Go? by Paul Showers. Explains how people create too much waste and how waste is now recycled and put into landfills. Let's-Read-and-Find-Out Science Stage 2. HarperCollins, 1994. [RL 3.1 IL K–3] (4615302 CC 4615301 PB)

•RL = Reading Level
•IL = Interest Level
Perfection Learning's catalog numbers are included for your ordering convenience.
PB indicates paperback. CC indicates Cover Craft. HB indicates hardback.

Glossary

commercialism
(kuh MER shuhl i zuhm) placing too much emphasis on making money

consumer (kuhn SOO mer) someone who buys goods or services

consumption
(kuhn SUHMP shuhn) the purchase and use of goods by **consumers** (see separate entry)

dump (dump) a pit in the ground where garbage is thrown

electronic waste
(ee lek TRON ik wayst) unwanted computers, monitors, televisions, audio equipment, and other devices that use electricity

extract (ek STRAKT) to obtain something from something else, usually by separating it out from other material

hazardous waste
(HAZ er duhs wayst) any waste substance that poses a risk to human or environmental health

incinerate (in SIN er ayt) to burn to ashes in a special furnace for destroying things by burning them

landfill (LAND fil) a pit lined with clay and plastic where **solid waste** is dumped (see separate entry)

leachate (LEE chayt) liquids that are produced in **dumps** as rain soaks through the layers of garbage (see separate entry)

methane (METH ayn) a gas that is produced when garbage breaks down and rots

municipal (myoo NIS uh puhl) relating to a town or city that has its own local government

natural resource (NA chur uhl REE sors) a natural material that can be used by people

organic (or GAN ik) derived from living things

packaging (PAK uh jing) what products are wrapped or boxed in

recycling (ree SEYEK ling) finding another use for a product or recovering materials from **solid waste** and using them to make new products (see separate entry)

solid waste (SAHL uhd wayst) trash or garbage produced by households, offices, and industries

source reduction (sors ree DUHK shuhn) creating less waste through decreased consumption or changes in product manufacturing or distribution

toxic (TAHK sik) poisonous or hazardous to one's health

waste (wayst) garbage or trash

waste stream (wayst streem) the amount of **waste** produced (see separate entry)

waste-to-energy plant (wayst too EN er jee plant) a plant that produces energy for electricity from incinerators that burn **solid waste** (see separate entry)

Index

Break of Dawn, 13
    Harrelson, Lowell, 13
chlorofluro carbons, 28
commercialism, 4
consumer, 5, 12, 15, 29
consumption, 4, 8, 12, 15
demanufacturing, 29
dump, 10, 23, 24, 25
early waste disposal
    Black Death (bubonic plague), 11
    Mayans, 10
    Middle Ages, 10–11
electronic waste, 26–29, 39, 41
    China, 26, 28
hazardous waste, 30, 31, 32, 36–37
    Bhopal, India, 32
    disposal of, 34–35
    household, 33–34
    Love Canal, 30–31
incinerate, 13, 21, 22, 43
Industrial Revolution, 11
landfills, 13, 14, 16, 23, 24, 25, 27, 28, 34,
        37, 43
leachate, 23, 24, 25
methane, 23, 24
municipal, 21
natural resources, 8
organic, 11

recycling, 7, 9, 11, 13, 16–20, 22, 26, 29,
        37–38, 39, 42–43
    aluminum, 12, 16, 18, 28
    "bottle bill", 37
    glass, 5, 6, 18, 20, 28
    metals, 5, 6, 16, 18, 20, 29
    paper, 9, 17, 18, 22, 43
    "pay-as-you-throw" programs, 38
    plastics, 6, 9, 12, 15, 19, 22, 29, 43
solid waste, 9, 21, 26
source reduction, 14–15, 40, 41
    packaging, 14–15, 19, 41
    pre-cycling, 38–39
strip mining, 5, 6
The Superfund Act, 36–37
toxic, 5, 6, 7, 32, 35, 36
waste, 4, 6, 7, 8, 9, 10, 11, 12, 14, 15, 27,
        38, 40, 42, 43
waste stream, 14, 16, 17, 36, 42, 43
waste-to-energy plants (incinerators), 21, 22